Inventions that will change the world in the next 100 years

Inventions that will change the world in the next 100 years

Sophie-Domingues-Montanari, PhD

DISCLAIMER

The author made every effort to ensure the accuracy and reliability of the information provided in this book, but it is ultimately the responsibility of the reader to exercise discretion and judgment when applying the contents herein.

Furthermore, the author hereby absolves themselves of any liability for the outcomes that may stem from the use, application, or reliance upon the information provided within this book. This disclaimer encompasses, but is not limited to, any potential loss, injury, or damage – whether direct, indirect, incidental, consequential, or punitive – resulting from actions taken based on the information herein.

By accessing and utilizing the contents of this book, readers acknowledge and agree to indemnify and hold harmless the author from any and all claims, demands, or liabilities that may arise.

COPYRIGHT STATEMENT

Copyright © Sophie Domingues-Montanari, 2024.

All rights reserved. No part of this publication may be reproduced, distributed, or transmitted in any form or by any means, including photocopying, recording, or other electronic or mechanical methods, without the prior written permission of the publisher, except in the case of brief quotations embodied in critical reviews and certain other noncommercial uses permitted by copyright law.

This book is protected under international copyright laws and treaties. Unauthorized reproduction, distribution, or transmission of this work is prohibited and may result in severe civil and criminal penalties.

TABLE OF CONTENTS

Prologue **13**

Powering Tomorrow: **15**

A Vision for the Energy Revolution **15**

   The Future of Energy *17*

   Energy Storage Solutions *29*

   Energy Distribution *35*

Driving Change: **41**

The Future of Transportation **41**

   Futuristic Transportation *43*

   Transportation Infrastructure *49*

   Sustainable Urban Mobility *53*

Healing Horizons: **59**

Navigating the Healthcare Evolution **59**

   Medical Technology *61*

   Artificial Intelligence *67*

   Regenerative Medicine *73*

Connecting Worlds: **77**

The Evolution of Communication 77

Communication Technology 79

Data Transmission 85

Augmented Reality 91

Eco-Evolution: 97

Advancing Environmental Innovation 97

Environmental Challenges 99

The role of biotechnology in environmental conservation and biodiversity preservation 105

Geoengineering solutions and their potential impact on climate change mitigation 109

Frontiers Beyond: 115

Charting the Course of Space Exploration and Colonization 115

Space Exploration Technology 117

Human Colonization of Mars 121

Space Tourism 125

Conclusion 131

Prologue

As we stand on the cusp of a new era marked by unprecedented technological advancement and innovation, the potential for transformative inventions to reshape our world has never been greater.

From advancements in renewable energy and transportation to breakthroughs in healthcare, communication, and space exploration, the inventions featured in this book represent the cutting edge of human ingenuity and creativity. Drawing on the latest scientific research, historical insights, and visionary thinking, we delve into the possibilities and implications of these inventions, offering a glimpse into a future where the boundaries of what is possible are continually pushed and redefined.

As we navigate the complex challenges and opportunities of the 21st century, it is essential to recognize the transformative potential of innovation in addressing pressing global issues such as climate change, healthcare disparities, and technological inequality. By harnessing the power of invention and imagination, we can unlock solutions to some of the most daunting challenges facing humanity and create a brighter, more sustainable future for generations to come.

In the pages that follow, you will encounter a diverse array of inventions and technologies that have the potential to shape the course of human history in the next 100 years. From renewable energy sources that promise to end our dependence on fossil fuels to medical breakthroughs that could cure diseases and prolong life, each invention offers a glimpse into a future where the impossible becomes possible and the unimaginable becomes reality.

Powering Tomorrow:

A Vision for the Energy Revolution

The Future of Energy

Nuclear Fusion: Unlocking the Power of the Stars

Nuclear fusion, the process that powers the sun and stars, holds the key to virtually limitless, clean energy. Unlike nuclear fission, which splits atoms to release energy, fusion involves fusing atomic nuclei to generate massive amounts of energy. While scientists have been striving to replicate this process on Earth for decades, significant challenges remain. However, recent breakthroughs in fusion research, such as the development of high-temperature superconductors and advanced confinement techniques, offer renewed hope.

One promising approach is magnetic confinement fusion, exemplified by projects like ITER (International Thermonuclear Experimental Reactor). ITER, a collaborative effort involving 35 nations, aims to demonstrate the feasibility of fusion as a practical energy source. By confining superheated plasma within a magnetic field, ITER hopes to achieve sustained fusion reactions, paving the way for commercial fusion power plants.

Meanwhile, alternative approaches like inertial confinement fusion, which uses high-energy lasers to compress and heat fusion fuel, are also being explored. While challenges remain, including the need for efficient energy capture and materials that can withstand the intense conditions inside a fusion reactor, the potential benefits of fusion energy are immense: abundant fuel, minimal waste, and no greenhouse gas emissions.

Resources

1. **ITER** (International Thermonuclear Experimental Reactor) - The official website of the ITER project, which is one of the most significant nuclear fusion experiments in the world.

    https://www.iter.org/

2. **National Ignition Facility** (NIF) - Operated by Lawrence Livermore National Laboratory, NIF conducts experiments in fusion ignition and energy production.

    https://lasers.llnl.gov/

3. **European Fusion Development Agreement** (EFDA) - The European initiative for coordinating fusion research activities in Europe.

    https://euro-fusion.org/

4. **General Atomics Fusion Energy Research** - General Atomics is a private company heavily involved in fusion energy research.

    https://www.ga.com/magnetic-fusion/

5. **Princeton Plasma Physics Laboratory** (PPPL) - A U.S. Department of Energy national laboratory dedicated to plasma physics and nuclear fusion research.

    https://www.pppl.gov/

6. **Max Planck Institute for Plasma Physics** (IPP) - A German research institute focusing on plasma physics and fusion research.

    https://www.ipp.mpg.de/en

7. **National Institute for Fusion Science** (NIFS) - A Japanese research institute dedicated to fusion science and related technologies.

    https://www.nifs.ac.jp/en/

8. **Fusion for Energy** (F4E) - The European Union's organization responsible for Europe's contribution to ITER.

   https://fusionforenergy.europa.eu/

9. **UK Atomic Energy Authority** (UKAEA) - Conducts fusion research at the Culham Centre for Fusion Energy.

   https://ccfe.ukaea.uk/

10. **KSTAR** (Korea Superconducting Tokamak Advanced Research) - South Korea's national fusion research institute.

    https://www.kfe.re.kr/

Advanced Solar Technologies: Harvesting the Power of the Sun

The sun, a boundless source of energy, has the potential to meet humanity's needs many times over. While traditional solar photovoltaic (PV) panels have made significant strides in efficiency and affordability, next-generation solar technologies promise even greater gains.

Perovskite solar cells, for example, offer the potential for high efficiency at low cost. Made from a class of materials with a unique crystal structure, perovskite solar cells can be manufactured using inexpensive processes, making solar energy more accessible to a wider range of users. With ongoing research focused on enhancing stability and scalability, perovskite solar cells could revolutionize the solar industry in the coming years.

Another promising area of innovation is tandem solar cells, which combine multiple materials with complementary absorption properties to achieve higher efficiency. By stacking layers of different semiconductors, tandem solar cells can capture a broader spectrum of sunlight, increasing overall energy conversion efficiency. With advancements in materials science and manufacturing techniques, tandem solar cells have the potential to surpass the efficiency limits of traditional silicon-based solar cells.

Resources

1. **National Renewable Energy Laboratory (NREL)** - NREL conducts research on various renewable energy technologies, including advanced solar technologies.

    https://www.nrel.gov/

2. **Solar Energy Research Institute of Singapore (SERIS)** - SERIS focuses on solar energy research, including advanced solar technologies and materials.

   https://www.seris.sg/

3. **Fraunhofer Institute for Solar Energy Systems (ISE)** - Fraunhofer ISE is a leading research institute in the field of solar energy, conducting research on advanced solar technologies.

   https://www.ise.fraunhofer.de/

4. **Solar Energy Research Center (SERC)** - SERC is a research center at the University of Texas at Austin dedicated to advancing solar energy technologies.

   https://serc.utexas.edu/

5. **Stanford Solar Energy Lab** - Stanford University's Solar Energy Lab conducts research on various aspects of solar energy, including advanced technologies.

   https://web.stanford.edu/group/solar/

6. **Australian National University (ANU) Solar Thermal Group** - ANU's Solar Thermal Group conducts research on advanced solar thermal technologies.

   https://solar.anu.edu.au/

7. **Solar Power and Chemical Energy Systems (SolarPACES)** - SolarPACES is a collaborative research program focused on advancing solar thermal energy technologies.

   https://www.solarpaces.org/

8. **European Solar Thermal Electricity Association (ESTELA)** - ESTELA represents the solar thermal electricity industry in Europe and provides information on research in the field.

   https://www.estelasolar.eu/

9. **Advanced Photovoltaics Concepts Center (APCC)** - APCC is a research center at the University of Michigan focused on advancing photovoltaic technologies.

   https://apcc.umich.edu/

10. **Solar Energy Industries Association (SEIA)** - SEIA represents the solar industry in the United States and provides information on research and developments in advanced solar technologies.

    https://www.seia.org/

Space-Based Solar Power: A Beacon of Hope from the Heavens

Imagine a world where clean, abundant solar energy is available around the clock, regardless of weather conditions or geographic location. Space-based solar power (SBSP) offers precisely that possibility. By capturing sunlight in space and transmitting it to Earth via microwave or laser beams, SBSP has the potential to revolutionize the way we harness solar energy.

The concept of SBSP dates back to the 1960s, when visionary scientists and engineers proposed the idea of placing solar power satellites in geostationary orbit. These satellites would be equipped with large solar arrays to capture sunlight, which would then be converted into microwave or laser beams and transmitted to receiving stations on Earth. Unlike ground-based solar installations, SBSP satellites would be unaffected by atmospheric interference or the day-night cycle, ensuring continuous energy delivery.

While the technical challenges of SBSP are formidable, including the development of lightweight, high-efficiency solar panels and efficient wireless power transmission systems, recent advances in space technology and materials science have renewed interest in this ambitious concept. Companies like SpaceX and Northrop Grumman are exploring the feasibility of SBSP, driven by the prospect of virtually unlimited clean energy and the potential to address energy needs on a global scale.

Resources

1. **National Space Society (NSS)** - NSS advocates for space exploration and development, including research on SBSP.

    https://space.nss.org/

2. **Space Solar Power Institute (SSPI)** - SSPI is dedicated to advancing research and education on space-based solar power systems.

   https://spacesolarpower.org/

3. **NASA - Space-Based Solar Power Project** - NASA conducts research on various space technologies, including SBSP.

   https://www.nasa.gov/directorates/spacetech/strg/Space-Based_Solar_Power

4. **International Academy of Astronautics (IAA)** - Study Group on Space Solar Power - IAA's Study Group on Space Solar Power focuses on advancing research in this field.

   https://iaaweb.org/content/view/562/847/

5. **The Planetary Society** - The Planetary Society explores space-related topics and advocates for space-based technologies like SBSP.

   https://www.planetary.org/

6. **European Space Agency (ESA)** - Space-Based Solar Power - ESA conducts research and studies on various space technologies, including SBSP.

   https://www.esa.int/Specials/Space-based_Solar_Power/index.html

7. **Defense Advanced Research Projects Agency (DARPA)** - Space-Based Solar Power - DARPA explores innovative technologies, including SBSP, for defense and national security purposes.

   https://www.darpa.mil/program/space-based-solar-power

8. **Japan Aerospace Exploration Agency (JAXA)** - Space Solar Power Systems - JAXA conducts research on various space technologies, including SBSP systems.

   https://global.jaxa.jp/projects/sat/sps/index.html

9. **The Space Frontier Foundation** - The Space Frontier Foundation promotes the development of space technologies, including SBSP.

   https://spacefrontier.org/

10. **SpaceWorks Engineering - Space Solar Power** - SpaceWorks Engineering conducts research and analysis on space-based technologies, including SBSP.

    https://www.spaceworks.aero/space-solar-power

Energy Storage Solutions

Energy storage is not a novel concept. Throughout history, civilizations have harnessed the power of stored energy in various forms, from ancient water wheels to medieval windmills. However, modern technology has propelled us into an era where storage solutions are more crucial than ever.

One of the most promising advancements in energy storage lies in battery technology. Batteries have come a long way since Alessandro Volta's voltaic pile in 1800, evolving from lead-acid to lithium-ion batteries that power our smartphones and electric vehicles today. But the real game-changer is the development of next-generation batteries with higher energy density, faster charging capabilities, and longer lifespans.

Among these, solid-state batteries are generating significant buzz. Unlike conventional lithium-ion batteries, which use liquid electrolytes, solid-state batteries employ solid electrolytes, offering improved safety and energy efficiency. Companies like Toyota and QuantumScape are racing to commercialize solid-state batteries for electric vehicles, promising longer ranges and shorter charging times.

Another promising avenue is flow battery technology, which separates energy storage from power output. Flow batteries utilize electrolyte solutions stored in external tanks, allowing for scalable and customizable energy storage capacities. This flexibility makes them ideal for grid-level applications, where demand fluctuations are common.

But perhaps the most intriguing prospect on the horizon is the concept of gravity-based energy storage. Gravity storage systems harness the potential energy of elevated weights or water reservoirs, releasing it as electricity when needed. Projects like the Gravity Power Module in Nevada aim to leverage this principle to store renewable energy and provide grid stability.

Beyond batteries, other innovative solutions are emerging. Power-to-gas technology converts excess renewable energy into hydrogen or synthetic natural gas, which can be stored and used for heat, transportation, or electricity generation when renewable sources are unavailable. Similarly, compressed air energy storage utilizes surplus energy to compress air into underground reservoirs, which is then released to drive turbines during peak demand.

The role of energy storage solutions in renewable energy adoption cannot be overstated. They serve as the linchpin for overcoming the intermittency of solar and wind power, enabling a smoother transition to a carbon-neutral energy landscape. By storing surplus energy when production exceeds demand and releasing it when needed, these technologies ensure grid stability and reliability, paving the way for a more resilient and sustainable energy infrastructure.

Moreover, energy storage solutions hold the key to unlocking the full potential of renewable resources. By mitigating the variability inherent in solar and wind energy, they facilitate their integration into existing power grids, reducing reliance on fossil fuels and greenhouse gas emissions. This not only addresses the urgent need to combat climate change but also fosters energy independence and security.

The economic implications of energy storage are equally significant. As the cost of renewable energy continues to decline, coupled with advancements in storage technology, the business case for clean energy becomes increasingly compelling. Energy storage systems offer opportunities for arbitrage, allowing operators to buy low-cost electricity when demand is low and sell it at higher prices during peak hours. This maximizes the value of renewable energy assets and accelerates their deployment worldwide.

Furthermore, energy storage solutions open new avenues for innovation and entrepreneurship. Startups and research institutions are actively exploring novel materials, designs, and applications to enhance the performance and affordability of storage technologies. This vibrant ecosystem fosters collaboration and competition, driving continuous improvement and driving down costs.

However, challenges remain on the path to widespread adoption of energy storage solutions. Technical hurdles such as limited energy density, efficiency losses, and material scarcity must be addressed to unlock the full potential of these technologies. Additionally, regulatory frameworks and market structures may need to evolve to incentivize investment in energy storage infrastructure and ensure fair competition in the energy market.

Resources

1. **Energy Storage Association (ESA)** - ESA is the national trade association dedicated to energy storage, advocating for policies and regulations that support its growth.
   https://energystorage.org/
2. **International Renewable Energy Agency (IRENA)** - Energy Storage - IRENA provides resources and reports

on energy storage technologies and their integration with renewable energy systems.

https://www.irena.org/energytransition/technologies/Energy-Storage

3. **U.S. Department of Energy (DOE)** - Energy Storage - DOE conducts research and development on energy storage technologies and provides information on funding opportunities and projects.

https://www.energy.gov/science-innovation/energy-storage

4. **Electric Power Research Institute (EPRI)** - Energy Storage - EPRI conducts research on various aspects of energy storage, including grid integration and technology advancements.

https://www.epri.com/research/technical-areas/energy-storage

5. **Clean Energy Group (CEG)** - CEG works on advancing innovative energy storage solutions for clean energy technologies, particularly for low-income and disadvantaged communities.

https://www.cleanegroup.org/

6. **Rocky Mountain Institute (RMI)** - Energy Storage - RMI conducts research and provides insights on energy storage technologies and their role in the transition to a clean energy future.

https://rmi.org/our-work/electricity/energy-storage/

7. **Energy Storage News** - Energy Storage News provides news, analysis, and market insights on energy storage technologies and industry developments.

https://www.energy-storage.news/

8. **BloombergNEF (BNEF)** - Energy Storage - BNEF offers research and analysis on energy storage markets, technologies, and investment trends.

   https://about.bnef.com/energy-storage/

9. **California Energy Storage Alliance (CESA)** - CESA promotes the adoption of energy storage technologies in California through advocacy, education, and policy development.

   https://www.storagealliance.org/

10. **Energy Storage Hub (ESH)** - ESH is an online platform that provides information and resources on energy storage technologies, applications, and market trends.

    https://www.energystoragehub.org/

Energy Distribution

Energy distribution has come a long way since the early days of centralized power plants and simple transmission lines. Historically, electricity was generated in large, remote facilities and transmitted over long distances to urban centers via high-voltage transmission lines. However, this traditional model posed several challenges, including transmission losses, grid congestion, and vulnerability to disruptions.

Enter the era of smart grids - an evolution of the traditional electricity grid empowered by digital technologies and real-time data analytics. At its core, a smart grid is an integrated system that monitors, controls, and optimizes the generation, transmission, and distribution of electricity, enabling more efficient, reliable, and sustainable energy delivery.

One of the key advancements in energy distribution is the deployment of advanced sensors and monitoring devices throughout the grid infrastructure. These sensors collect real-time data on variables such as voltage, current, and power flow, providing grid operators with unprecedented visibility into the state of the system. By continuously monitoring grid conditions, operators can detect and respond to disturbances more quickly, minimizing the risk of outages and optimizing energy flow.

Another critical component of smart grids is the implementation of advanced communication and control systems. These systems enable bidirectional communication between grid components, allowing for dynamic adjustments in response to changing conditions. For example, automated switches and reclosers can isolate faulty sections of the grid and reroute power to minimize disruption to customers. Similarly, demand response programs enable utilities to adjust electricity consumption in real-time, leveraging price signals to incentivize load-shifting and reduce peak demand.

Renewable energy integration is another area where smart grid technologies are making a significant impact. As solar and wind power continue to proliferate, grid operators face the challenge of balancing intermittent generation with fluctuating demand. Smart grids enable seamless integration of renewable energy sources by providing real-time forecasting, grid-friendly inverters, and energy storage solutions. By dynamically managing generation and consumption, smart grids maximize the utilization of renewable resources while maintaining grid stability and reliability.

Furthermore, smart grids empower consumers to actively participate in the energy system through demand-side management and distributed generation. Smart meters and home energy management systems allow consumers to monitor their electricity usage in real-time and adjust consumption patterns accordingly. This not only helps reduce energy bills but also contributes to overall grid efficiency by flattening demand curves and reducing peak load.

The benefits of smart grid technologies extend beyond operational efficiency to encompass environmental sustainability and resilience. By optimizing energy flow and reducing wastage, smart grids contribute to lower greenhouse gas emissions and mitigate the impact of climate change. Moreover, the decentralized nature of smart grids makes them inherently more resilient to disruptions, whether caused by natural disasters or cyber-attacks.

However, the transition to smart grids is not without challenges. Legacy infrastructure, regulatory barriers, and cybersecurity concerns pose significant hurdles to widespread adoption. Additionally, the complexity and cost of implementing smart grid technologies may deter some utilities from embracing these innovations.

Despite these challenges, the momentum behind smart grids continues to grow as governments, utilities, and technology providers recognize the potential for transformative change. Initiatives such as the Smart Grid Investment Grant Program in the United States and the European Union's Smart Grids Task Force are driving investment and collaboration in smart grid development.

Resources

1. **Electric Power Research Institute (EPRI)** - EPRI conducts research on various aspects of electricity generation, transmission, and distribution.

    https://www.epri.com/

2. **Edison Electric Institute (EEI)** - EEI represents investor-owned electric companies in the United States and provides information on energy distribution and grid modernization.

    https://www.eei.org/

3. **Smart Electric Power Alliance (SEPA)** - SEPA focuses on the integration of renewable energy, distributed energy resources, and grid modernization.

    https://sepapower.org/

4. **North American Electric Reliability Corporation (NERC)** - NERC ensures the reliability and security of the North

American bulk power system, including energy distribution networks.

https://www.nerc.com/

5. **Gridwise Alliance** - Gridwise Alliance advocates for modernizing the electric grid and improving energy distribution infrastructure.

https://www.gridwise.org/

6. **International Energy Agency (IEA)** - Electricity Security Analysis Center (ESAC) - IEA's ESAC provides analysis and insights on electricity security, including energy distribution.

https://www.iea.org/topics/electricity-security-analysis-centre

7. **American Public Power Association (APPA)** - APPA represents publicly-owned electric utilities and provides resources on energy distribution and utility management.

https://www.publicpower.org/

8. **Institute of Electrical and Electronics Engineers (IEEE)** - Power & Energy Society (PES) - IEEE PES focuses on advancements in power and energy systems, including energy distribution technologies.

https://www.ieee-pes.org/

9. **European Network of Transmission System Operators for Electricity (ENTSO-E)** - ENTSO-E coordinates the operation of the European electricity transmission system, including energy distribution.

https://www.entsoe.eu/

10. **Electricity Distributors Association (EDA)** - EDA represents Ontario's local electricity distribution

companies and provides information on energy distribution in the province.

https://www.eda-on.ca/

Driving Change:

The Future of Transportation

Futuristic Transportation

At the forefront of the transportation revolution are futuristic methods that challenge the conventional notions of speed, efficiency, and sustainability. Among these, the hyperloop stands out as a vision of high-speed travel propelled by magnetic levitation and low-pressure tubes. Conceived by Elon Musk in 2013, the hyperloop concept envisions pod-like capsules traveling at speeds exceeding 700 miles per hour, drastically reducing travel times between cities.

The key to the hyperloop's potential lies in its utilization of vacuum-sealed tubes and electromagnetic propulsion to eliminate air resistance and friction, enabling near frictionless travel. While the hyperloop remains largely conceptual, several companies, including Virgin Hyperloop and SpaceX, are actively pursuing prototype development and feasibility studies. If successful, the hyperloop could revolutionize long-distance travel, offering a faster, more sustainable alternative to air travel and conventional high-speed rail.

In addition to the hyperloop, flying cars represent another futuristic transportation method capturing the imagination of engineers and innovators worldwide. The concept of flying cars has long been a staple of science fiction, but recent advancements in electric propulsion, autonomy, and urban air mobility are bringing this futuristic vision closer to reality.

Electric Vertical Takeoff and Landing (eVTOL) aircraft, also known as flying taxis, promise to revolutionize urban transportation by taking to the skies and bypassing traffic congestion below. Companies like Uber, Boeing, and Volocopter are investing heavily in eVTOL technology, aiming to create a new paradigm of on-demand, point-to-point aerial transportation.

The allure of flying cars lies not only in their potential to alleviate traffic congestion but also in their environmental benefits. Electric propulsion eliminates harmful emissions associated with traditional combustion engines, reducing air pollution and greenhouse gas emissions in urban environments. Moreover, the vertical takeoff and landing capabilities of eVTOL aircraft enable them to operate from existing infrastructure, such as helipads and rooftops, minimizing the need for costly new infrastructure development.

Meanwhile, autonomous vehicles represent yet another frontier in transportation innovation, promising to redefine the way we perceive and interact with automobiles. Autonomous vehicles, or self-driving cars, leverage advanced sensors, artificial intelligence, and real-time data processing to navigate roads and highways without human intervention.

The development of autonomous vehicles has been fueled by major advancements in sensor technology, particularly lidar (light detection and ranging), radar, and cameras, which enable vehicles to perceive their surroundings with unprecedented accuracy and reliability. Combined with sophisticated algorithms and machine learning algorithms, autonomous vehicles can analyze vast amounts of data in real-time to make split-second decisions and navigate complex traffic scenarios safely.

While fully autonomous vehicles have yet to achieve widespread deployment, significant progress has been made in testing and development by companies like Waymo, Tesla, and General Motors. Pilot programs and demonstrations in cities around the world are showcasing the potential of autonomous vehicles to enhance safety, improve mobility access, and reduce traffic accidents caused by human error.

Beyond personal transportation, autonomous vehicles hold promise for revolutionizing freight logistics, public transit, and last-mile delivery services. Automated trucks and drones offer opportunities for more efficient and cost-effective freight transport, while autonomous buses and shuttles promise to enhance mobility options for urban residents and reduce reliance on private car ownership.

However, the adoption of autonomous vehicles also raises complex ethical, regulatory, and societal questions. Concerns about safety, liability, privacy, and job displacement must be addressed to ensure the responsible integration of autonomous technologies into our transportation infrastructure. Moreover, equitable access to autonomous mobility solutions must be prioritized to ensure that all communities benefit from the potential of this transformative technology.

Resources

1. **Hyperloop Transportation Technologies (HTT)** - HTT is developing the Hyperloop, a futuristic mode of transportation that involves high-speed travel in low-pressure tubes.

   https://www.hyperlooptt.com/

2. **Virgin Hyperloop** - Virgin Hyperloop is another company working on developing the Hyperloop technology for ultra-fast transportation.

   https://virginhyperloop.com/

3. **Tesla** - Tesla, led by Elon Musk, is known for its advancements in electric vehicles and autonomous

driving technology, which are shaping the future of transportation.

https://www.tesla.com/

4. **Uber Elevate** - Uber Elevate is exploring the concept of urban air mobility through electric vertical takeoff and landing (eVTOL) aircraft for future transportation solutions.

   https://www.uber.com/us/en/elevate/

5. **The Boring Company** - Founded by Elon Musk, The Boring Company aims to revolutionize transportation by constructing underground tunnels for high-speed transit systems.

   https://www.boringcompany.com/

6. **AeroMobil** - AeroMobil is developing flying cars, blending the convenience of road vehicles with the freedom of flight for futuristic transportation solutions.

   https://www.aeromobil.com/

7. **Terrafugia** - Terrafugia is working on the development of flying cars and other innovative personal air vehicles.

   https://terrafugia.com/

8. **NASA** - Advanced Air Mobility (AAM) - NASA's AAM initiative explores advanced concepts for air transportation, including electric aircraft and urban air mobility solutions.

   https://www.nasa.gov/aam

9. **BMW Group - BMW i Ventures** - BMW i Ventures invests in innovative transportation startups, focusing on technologies that shape the future of mobility.

   https://www.bmwiventures.com/

10. **Toyota Research Institute (TRI)** - TRI conducts research on artificial intelligence, robotics, and autonomous vehicles to develop future transportation solutions.

    https://www.tri.global/

Transportation Infrastructure

In recent years, two key trends have emerged that are revolutionizing transportation: electrification and automation. These transformative technologies hold the potential to not only redefine how we move people and goods but also to address pressing challenges such as climate change, congestion, and road safety.

First, let's delve into the impact of electrification on transportation infrastructure. The shift towards electrification involves replacing traditional fossil fuel-powered vehicles with electric vehicles (EVs) that run on electricity stored in batteries or fuel cells. This transition is driven by the urgent need to reduce greenhouse gas emissions and mitigate the impact of climate change.

Electric vehicles offer several advantages over their gasoline-powered counterparts. They produce zero tailpipe emissions, thereby improving air quality and reducing harmful pollutants such as nitrogen oxides and particulate matter. Additionally, EVs are quieter and require less maintenance than internal combustion engine vehicles, leading to reduced noise pollution and lower operating costs.

The widespread adoption of electric vehicles is also driving changes in transportation infrastructure. Charging infrastructure, including public charging stations and fast-charging networks, is expanding rapidly to support the growing fleet of EVs on the road. Governments, utilities, and private companies are investing heavily in charging infrastructure to address range anxiety and facilitate long-distance travel for EV owners.

Moreover, the electrification of transportation is spurring innovation in energy storage and grid integration. Vehicle-to-grid (V2G) technology enables EVs to serve as mobile energy storage units, providing grid services such as peak shaving, load balancing, and renewable energy integration. This bi-directional flow of electricity between vehicles and the grid has the potential to enhance grid stability, increase renewable energy penetration, and reduce the need for costly infrastructure upgrades.

The impact of automation on transportation infrastructure is equally profound. Automation refers to the use of advanced technologies such as artificial intelligence, sensors, and robotics to enable vehicles to operate without direct human intervention. This includes both fully autonomous vehicles (AVs) and advanced driver assistance systems (ADAS) that enhance the capabilities of human drivers.

Automation has the potential to revolutionize transportation infrastructure in several ways. One of the most significant impacts is the optimization of traffic flow and congestion management. Autonomous vehicles can communicate with each other and with traffic management systems to coordinate their movements, reducing bottlenecks, and improving overall traffic efficiency.

Furthermore, automation has the potential to enhance safety and reduce traffic accidents. Human error is a leading cause of road accidents, accounting for the majority of crashes worldwide. Autonomous vehicles, equipped with advanced sensors and algorithms, have the potential to eliminate human error and drastically reduce the incidence of accidents.

Additionally, automation has the potential to increase accessibility and mobility for people who are unable to drive due to age, disability, or other factors. Autonomous ride-hailing services and on-demand transportation solutions promise to provide convenient and affordable mobility options for all members of society, regardless of their ability to drive.

However, the widespread adoption of automation also raises important questions about infrastructure compatibility, cybersecurity, and ethical considerations. Existing transportation infrastructure, designed primarily for human-operated vehicles, may need to be upgraded or modified to accommodate autonomous vehicles effectively. Additionally, concerns about data privacy, cybersecurity, and liability must be addressed to ensure the safe and responsible deployment of automated transportation systems.

Resources

1. **Electric Power Research Institute (EPRI)** - EPRI conducts research on electrification technologies for transportation, including electric vehicles and charging infrastructure.

    https://www.epri.com/

2. **International Council on Clean Transportation (ICCT)** - ICCT conducts research on clean transportation technologies, including electrification and automation.

    https://www.theicct.org/

3. **Smart Transportation Alliance (STA)** - STA promotes research and innovation in smart transportation solutions, including electrification and automation.

    https://smart-transportation.org/

4. **National Renewable Energy Laboratory (NREL)** - Transportation & Mobility Research - NREL conducts research on renewable energy technologies for transportation, including electrification.

   https://www.nrel.gov/transportation/

5. **Institute of Transportation Studies (ITS)** - ITS focuses on research and education in transportation, including electrification and automation.

   https://its.ucdavis.edu/

6. **Center for Urban Transportation Research (CUTR)** - CUTR conducts research on urban transportation systems, including electrification and automation.

   https://www.cutr.usf.edu/

7. **Urban Mobility & Automated Driving (UMAD)** - UMAD is a research group at the Technical University of Munich focused on urban mobility and automation.

   https://www.umad.tech/

8. **Center for Automotive Research (CAR)** - CAR conducts research on automotive technologies, including electrification and automation.

   https://car.osu.edu/

9. **Transportation Electrification Research Center (TERC)** - TERC conducts research on transportation electrification technologies and infrastructure.

   https://terc.ucdavis.edu/

10. **Center for Transportation and the Environment (CTE)** - CTE conducts research on clean transportation technologies, including electrification and automation.

    https://www.cte.tv/

Sustainable Urban Mobility

At the heart of sustainable urban mobility lies the concept of multimodal transportation. Rather than relying solely on personal vehicles or traditional public transit, multimodal systems integrate various modes of transportation such as walking, cycling, public transit, and shared mobility services into a seamless and interconnected network. This approach not only reduces reliance on single-occupancy vehicles but also promotes healthier and more sustainable modes of travel.

One of the most visible manifestations of sustainable urban mobility is the resurgence of cycling as a viable mode of transportation. Cycling offers numerous benefits, including zero emissions, reduced congestion, and improved public health. Cities around the world are investing in cycling infrastructure such as bike lanes, bike-sharing programs, and secure bike parking to encourage more people to choose cycling for their daily commute.

Similarly, walking is gaining recognition as a fundamental mode of urban transportation. Pedestrian-friendly design principles, such as wider sidewalks, pedestrian crossings, and traffic-calming measures, enhance safety and accessibility for pedestrians. Additionally, mixed-use development and compact urban design promote walkability by bringing essential amenities within walking distance of residential areas, reducing the need for car trips.

Public transit remains a cornerstone of sustainable urban mobility, providing efficient and affordable transportation options for millions of city dwellers worldwide. Investments in high-capacity transit systems, such as buses, trams, and subways, improve mobility and accessibility while reducing greenhouse gas emissions and air pollution. Furthermore, advancements in electrification and automation are transforming public transit fleets, with electric buses and autonomous shuttles offering cleaner and more efficient transportation solutions.

Shared mobility services, such as ride-hailing, car-sharing, and bike-sharing, are also playing an increasingly important role in sustainable urban mobility. These services leverage technology to optimize vehicle utilization, reduce the need for private car ownership, and provide flexible and on-demand transportation options. By promoting shared mobility, cities can alleviate congestion, reduce parking demand, and enhance access to transportation for underserved communities.

The implications of sustainable urban mobility extend beyond transportation to encompass broader urban planning and design considerations. Sustainable cities prioritize compact, mixed-use development that minimizes sprawl and encourages walking, cycling, and public transit use. They prioritize green spaces, pedestrian-friendly streetscapes, and active transportation infrastructure to enhance quality of life and promote community health and well-being.

Furthermore, sustainable urban mobility strategies prioritize equity and social inclusion, ensuring that all members of society have access to safe, affordable, and reliable transportation options. This includes addressing barriers such as income inequality, geographic disparities, and transportation deserts that disproportionately affect marginalized communities. By prioritizing equity, cities can create more inclusive and resilient urban environments that benefit everyone.

However, the transition to sustainable urban mobility is not without challenges. Funding constraints, regulatory barriers, and political opposition can hinder progress and delay the implementation of sustainable transportation initiatives. Additionally, entrenched car-centric attitudes and behaviors may pose resistance to change, requiring concerted efforts to shift cultural norms and perceptions around transportation.

Nevertheless, the momentum behind sustainable urban mobility is growing, driven by a growing awareness of the urgent need to address climate change, improve air quality, and enhance quality of life in cities. Governments, businesses, and civil society organizations are increasingly recognizing the economic, environmental, and social benefits of investing in sustainable transportation infrastructure.

Resources

1. **International Association of Public Transport (UITP)** - UITP conducts research and provides resources on sustainable urban mobility, including public transit systems and policy.

    https://www.uitp.org/

2. **ICLEI - Local Governments for Sustainability** - ICLEI supports cities in implementing sustainable urban mobility solutions through research, policy guidance, and capacity building.

    https://www.iclei.org/

3. **Urban Transport Systems Laboratory (LUTS)** - LUTS is a research group at the Swiss Federal Institute of Technology in Lausanne (EPFL) focusing on sustainable urban mobility.

    https://luts.epfl.ch/

4. **Sustainable Urban Mobility Planning (SUMP) Platform** - The SUMP Platform provides guidance, tools, and research on sustainable urban mobility planning for cities across Europe.

   https://www.eltis.org/

5. **Institute for Transportation & Development Policy (ITDP)** - ITDP conducts research and advocacy on sustainable transportation solutions, including urban mobility.

   https://www.itdp.org/

6. **Transport Research and Innovation Portal (TRIP)** - TRIP provides access to research projects, publications, and resources on sustainable urban mobility funded by the European Commission.

   https://trimis.ec.europa.eu/

7. **Transportation Sustainability Research Center (TSRC)** - TSRC conducts interdisciplinary research on sustainable transportation, including urban mobility and policy analysis.

   https://tsrc.berkeley.edu/

8. **International Transport Forum (ITF)** - Decarbonising Transport Initiative - ITF's Decarbonising Transport Initiative focuses on research and policy analysis to promote sustainable urban mobility.

   https://www.itf-oecd.org/decarbonising-transport

9. **Urban Transport Group** - The Urban Transport Group is a UK-based organization that conducts research and advocacy on sustainable urban mobility solutions for cities.

   https://www.urbantransportgroup.org/

10. **Center for Sustainable Urban Development (CSUD)** - CSUD conducts research on sustainable urban development, including transportation and mobility, at Columbia University.

    https://csud.ei.columbia.edu/

Healing Horizons:

Navigating the Healthcare Evolution

Medical Technology

The realm of medical technology has undergone a remarkable transformation in recent decades, ushering in a new era of healthcare that is more personalized, precise, and effective than ever before. At the forefront of this revolution are breakthroughs in nanomedicine, personalized medicine, and gene editing, each offering unprecedented opportunities to diagnose, treat, and prevent a wide range of diseases and conditions.

Nanomedicine, the application of nanotechnology to medicine, represents one of the most exciting frontiers in healthcare innovation. At the nanoscale, materials exhibit unique properties that can be harnessed for targeted drug delivery, imaging, and diagnostics. Nanomedicine holds the promise of revolutionizing how we diagnose and treat diseases by enabling therapies that are more precise, potent, and less invasive than traditional approaches.

One of the most promising applications of nanomedicine is targeted drug delivery. Conventional drug formulations often suffer from poor bioavailability and off-target effects, leading to suboptimal therapeutic outcomes and adverse side effects. Nanoparticle-based drug delivery systems address these challenges by encapsulating drugs within nanoscale carriers that can be engineered to target specific cells or tissues in the body.

These nanocarriers can navigate through biological barriers, such as the blood-brain barrier or tumor microenvironment, to deliver therapeutic payloads directly to the site of action. This targeted approach minimizes systemic toxicity and maximizes drug efficacy, improving patient outcomes and quality of life.

Additionally, nanomedicine enables combination therapies, where multiple drugs or therapeutic agents are co-delivered to synergistically target complex diseases.

In addition to drug delivery, nanomedicine holds promise for advanced imaging and diagnostics. Nanoparticles can be engineered to enhance the contrast and sensitivity of imaging modalities such as magnetic resonance imaging (MRI), computed tomography (CT), and positron emission tomography (PET). These nanoparticle-based contrast agents enable earlier detection and more accurate staging of diseases, facilitating timely intervention and improving patient prognosis.

Personalized medicine, also known as precision medicine, represents another paradigm shift in healthcare, moving away from a one-size-fits-all approach towards treatments tailored to individual patients based on their unique genetic makeup, lifestyle, and environmental factors. Advances in genomics, proteomics, and data analytics have paved the way for personalized medicine, enabling healthcare providers to make more informed treatment decisions and optimize patient outcomes.

Central to personalized medicine is the concept of biomarker-driven therapy, where biomarkers such as genetic mutations, protein expression patterns, or metabolic signatures are used to guide treatment selection and dosing. By identifying patients who are most likely to respond to a particular therapy and predicting their risk of adverse reactions, personalized medicine maximizes treatment efficacy while minimizing the risk of unnecessary side effects.

One of the most notable successes of personalized medicine is the development of targeted cancer therapies. These therapies, which include small molecule inhibitors, monoclonal antibodies, and immunotherapies, are designed to selectively target cancer cells while sparing healthy tissues. By matching patients with the most effective therapies based on their tumor's genetic profile, personalized medicine has revolutionized cancer treatment and improved survival rates for many patients.

Another promising application of personalized medicine is pharmacogenomics, which studies how genetic variations influence an individual's response to drugs. By analyzing genetic variants that affect drug metabolism, efficacy, and toxicity, healthcare providers can optimize medication selection and dosing for each patient, minimizing the risk of adverse drug reactions and treatment failure.

Gene editing technologies, such as CRISPR-Cas9, have emerged as powerful tools for manipulating the genetic code with unprecedented precision and efficiency. CRISPR-Cas9 enables researchers to make targeted modifications to the genome by precisely cutting and editing DNA sequences. This technology has revolutionized basic research, allowing scientists to study gene function and disease mechanisms with unprecedented precision.

In addition to its research applications, gene editing holds immense promise for the treatment of genetic diseases and disorders. By correcting disease-causing mutations or introducing therapeutic genes into patient cells, gene editing therapies have the potential to cure a wide range of genetic disorders, including cystic fibrosis, sickle cell disease, and muscular dystrophy.

Furthermore, gene editing offers new possibilities for cancer immunotherapy, where immune cells are engineered to recognize and attack cancer cells more effectively. By modifying immune cells to express chimeric antigen receptors (CARs) or other tumor-targeting molecules, researchers can enhance the specificity and potency of immunotherapies, leading to better outcomes for cancer patients.

Despite these remarkable advances, challenges remain in translating breakthroughs in medical technology into clinical practice. Ethical considerations, regulatory frameworks, and safety concerns must be carefully addressed to ensure the responsible development and deployment of these technologies. Additionally, disparities in access to healthcare and genomic data privacy must be addressed to ensure that the benefits of personalized medicine and gene editing are equitably distributed.

Resources

1. **National Institutes of Health (NIH) - Nanomedicine Initiative** - NIH conducts research and supports initiatives focused on advancing nanomedicine technologies for healthcare applications.

    https://www.nih.gov/research-training/medical-research-initiatives/nanomedicine

2. **National Center for Advancing Translational Sciences (NCATS)** - NCATS supports research on translational science, including personalized medicine approaches to accelerate the development of new treatments and diagnostics.

    https://ncats.nih.gov/

3. **National Human Genome Research Institute (NHGRI) - Genomic Medicine Program** - NHGRI conducts research

on genomics and personalized medicine to understand the role of genetics in health and disease.

https://www.genome.gov/Genomic-Medicine

4. **European Medicines Agency (EMA) - Advanced Therapy Medicinal Products (ATMPs)** - EMA provides information and regulatory guidance on advanced therapies, including gene editing technologies used in medical treatments.

https://www.ema.europa.eu/en/human-regulatory/overview/advanced-therapy-medicinal-products-overview

5. **American Society of Gene & Cell Therapy (ASGCT)** - ASGCT is a professional organization focused on advancing research and clinical applications of gene and cell therapy.

https://asgct.org/

6. **International Society for Nanomedicine (ISN)** - ISN promotes research and collaboration in the field of nanomedicine, with a focus on its applications in healthcare.

https://www.isn.nagoya-u.ac.jp/

7. **Personalized Medicine Coalition (PMC)** - PMC advocates for personalized medicine approaches and provides resources on the latest developments in the field.

https://www.personalizedmedicinecoalition.org/

8. **The Journal of Gene Medicine** - This journal publishes research articles and reviews on gene therapy and related technologies.

https://onlinelibrary.wiley.com/journal/15212598

9. **Nature Nanotechnology** - Nature Nanotechnology publishes cutting-edge research on nanoscience and nanotechnology, including applications in medicine and healthcare.

   https://www.nature.com/nnano/

10. **Journal of Personalized Medicine** - This journal publishes research on personalized medicine, including genomics, biomarkers, and individualized treatment approaches.

    https://www.mdpi.com/journal/jpm

Artificial Intelligence

Artificial intelligence (AI) has emerged as a transformative force in healthcare, revolutionizing the way we diagnose, treat, and prevent diseases. In the quest for more personalized and efficient healthcare solutions, AI is paving the way towards a future where medical decisions are informed by data-driven insights and predictive analytics.

At its core, AI refers to the ability of computer systems to perform tasks that typically require human intelligence, such as understanding natural language, recognizing patterns, and making decisions. In healthcare, AI algorithms analyze vast amounts of patient data, including electronic health records, medical imaging, genomic sequences, and wearable sensor data, to extract meaningful insights and support clinical decision-making.

One of the most promising applications of AI in healthcare is medical imaging interpretation. Radiology, pathology, and other imaging-based specialties generate enormous volumes of medical images, which can be time-consuming and labor-intensive for human radiologists and pathologists to interpret accurately. AI algorithms, trained on large datasets of annotated images, can assist clinicians in detecting abnormalities, diagnosing diseases, and guiding treatment decisions with greater speed and accuracy.

For example, AI-powered image analysis systems have demonstrated remarkable performance in detecting breast cancer, lung nodules, and diabetic retinopathy from medical images. By leveraging deep learning algorithms and convolutional neural networks, these systems can identify subtle patterns and features that may be imperceptible to the human eye, enabling earlier detection and intervention for patients.

AI is also transforming clinical decision support by synthesizing vast amounts of clinical data and scientific literature to assist healthcare providers in making informed treatment decisions. Clinical decision support systems analyze patient data, including medical history, laboratory results, and treatment guidelines, to generate personalized treatment recommendations and alert clinicians to potential adverse events or drug interactions.

Moreover, AI-powered predictive analytics can forecast patient outcomes, identify at-risk populations, and optimize resource allocation in healthcare settings. Machine learning models trained on longitudinal patient data can predict the likelihood of hospital readmissions, patient deterioration, and disease progression, enabling proactive interventions and preventive care strategies.

In drug discovery and development, AI holds the promise of accelerating the discovery process, reducing costs, and increasing the success rate of new therapies. Traditional drug discovery methods are time-consuming, expensive, and often inefficient, with many promising drug candidates failing to reach clinical approval due to safety concerns or lack of efficacy.

AI algorithms, such as generative adversarial networks and reinforcement learning, can simulate molecular interactions, predict drug-target interactions, and design novel compounds with desired properties. By analyzing large datasets of chemical structures, biological assays, and clinical outcomes, AI-powered drug discovery platforms can identify potential drug candidates more efficiently and prioritize those with the highest likelihood of success.

Furthermore, AI-enabled clinical trials can streamline patient recruitment, improve trial design, and enhance data collection and analysis. By leveraging electronic health records, wearable devices, and real-world evidence, AI algorithms can identify eligible patients, stratify patient populations, and monitor trial progress in real-time, accelerating the development and approval of new therapies.

Despite its transformative potential, the widespread adoption of AI in healthcare is not without challenges. Concerns about data privacy, security, and bias must be addressed to ensure the ethical and responsible use of AI algorithms in clinical practice. Additionally, regulatory frameworks and reimbursement policies may need to evolve to accommodate AI-driven healthcare solutions and ensure patient safety and efficacy.

Moreover, the integration of AI into healthcare workflows requires investment in infrastructure, training, and collaboration between healthcare providers, technology companies, and regulatory agencies. Interdisciplinary approaches that combine expertise in medicine, computer science, and data analytics are essential for harnessing the full potential of AI to improve patient outcomes and advance medical science.

Resources

1. **National Institute of Health (NIH)** - National Institute of Biomedical Imaging and Bioengineering (NIBIB) - NIBIB conducts research on the application of artificial intelligence in biomedical imaging and healthcare.

    https://www.nibib.nih.gov/

2. **Healthcare Information and Management Systems Society (HIMSS)** - HIMSS provides resources and

insights on the adoption of artificial intelligence and other health information technologies in healthcare.

https://www.himss.org/

3. **American Medical Association (AMA)** - Artificial Intelligence in Healthcare - AMA offers information and guidance on the ethical and practical implications of artificial intelligence in healthcare.

   https://www.ama-assn.org/practice-management/digital/artificial-intelligence-health-care

4. **Journal of Artificial Intelligence in Medicine** - This journal publishes research articles and reviews on the applications of artificial intelligence and machine learning in medicine and healthcare.

   https://www.journals.elsevier.com/artificial-intelligence-in-medicine

5. **AI in Healthcare Summit** - The AI in Healthcare Summit provides a platform for researchers and practitioners to discuss the latest advancements and applications of artificial intelligence in healthcare.

   https://www.re-work.co/events/ai-in-healthcare-summit-boston-2022

6. **Stanford AIMI Center** - Stanford's Center for Artificial Intelligence in Medicine and Imaging conducts research on the development and deployment of AI technologies in healthcare.

   https://aimi.stanford.edu/

7. **MIT Critical Data** - MIT's Critical Data group conducts research on the application of artificial intelligence and data analytics in healthcare to improve patient outcomes.

   https://criticaldata.mit.edu/

8. **European Federation for Medical Informatics (EFMI)** - EFMI promotes research and education in medical informatics, including the use of artificial intelligence in healthcare.

   https://www.efmi.org/

9. **Artificial Intelligence Healthcare Conference** - This conference brings together experts from academia, industry, and healthcare to discuss the latest developments and challenges in AI-driven healthcare.

   https://aihealthcareconference.com/

10. **Artificial Intelligence in Medicine (AIME) Conference** - AIME is a conference series focused on artificial intelligence applications in medicine, featuring research presentations and workshops.

    http://aime20.aimedicine.info/

Regenerative Medicine

At the heart of regenerative medicine lies the concept of stem cells - unique cells with the remarkable ability to differentiate into various cell types and regenerate damaged tissues. Stem cells can be found in various sources, including embryonic tissue, adult tissues such as bone marrow and adipose tissue, and induced pluripotent stem cells (iPSCs) derived from reprogrammed adult cells.

One of the most promising applications of stem cells is in the field of organ transplantation. Organ failure is a leading cause of morbidity and mortality worldwide, with millions of patients awaiting life-saving transplants. However, the demand for donor organs far outstrips the supply, leading to long waiting lists and a shortage of suitable donor organs.

Regenerative medicine offers a potential solution to this crisis by enabling the generation of patient-specific organs and tissues in the laboratory. Scientists are exploring various approaches to organ regeneration, including the use of stem cells to grow functional organs, such as kidneys, livers, hearts, and lungs, in vitro. These bioengineered organs could be custom-tailored to match the patient's immune profile, reducing the risk of rejection and eliminating the need for lifelong immunosuppressive therapy.

Tissue engineering is another key component of regenerative medicine, focusing on the development of biomimetic scaffolds and bioactive materials to support cell growth and tissue regeneration. These scaffolds serve as three-dimensional templates that mimic the native extracellular matrix of tissues, providing structural support and biochemical cues to guide cell proliferation and differentiation.

In recent years, significant progress has been made in the field of tissue engineering, with researchers successfully engineering complex tissues and organs, including skin, cartilage, bone, and blood vessels, for transplantation and regenerative therapies. These bioengineered tissues hold the potential to repair damaged tissues, restore function, and improve quality of life for patients suffering from a wide range of conditions, including traumatic injuries, degenerative diseases, and congenital defects.

Moreover, advances in biomaterials and tissue engineering techniques have enabled the development of organ-on-a-chip and body-on-a-chip platforms for drug screening, disease modeling, and personalized medicine. These microfluidic devices replicate the physiological microenvironment of tissues and organs, allowing researchers to study disease mechanisms, test potential therapeutics, and predict drug responses with greater accuracy and efficiency than traditional cell culture models.

Despite these advancements, significant challenges remain on the path to widespread adoption of regenerative medicine technologies. Issues such as immune rejection, tumorigenicity, and vascularization must be overcome to ensure the safety and efficacy of bioengineered tissues and organs. Additionally, regulatory approval, scalability, and cost-effectiveness are critical factors that must be addressed to enable the translation of regenerative medicine from the laboratory to the clinic.

Resources

1. **National Institutes of Health (NIH) - Stem Cell Information** - NIH provides comprehensive information on stem cells, including their biology, potential applications, and ethical considerations.

    https://stemcells.nih.gov/

2. **International Society for Stem Cell Research (ISSCR)** - ISSCR is a professional organization dedicated to promoting stem cell research and its applications in regenerative medicine.

    https://www.isscr.org/

3. **California Institute for Regenerative Medicine (CIRM)** - CIRM funds stem cell research in California and provides resources and information on the latest developments in regenerative medicine.

    https://www.cirm.ca.gov/

4. **EuroStemCell** - EuroStemCell provides information and resources on stem cell research and regenerative medicine for the public and researchers in Europe.

    https://www.eurostemcell.org/

5. **The Stem Cell Network (SCN)** - SCN is a Canadian research network focused on advancing stem cell research and its translation into clinical applications.

    https://stemcellnetwork.ca/

6. **Stem Cell Reports** - Stem Cell Reports is an open-access journal that publishes research articles and reviews on stem cell biology and regenerative medicine.

    https://www.cell.com/stem-cell-reports/home

7. **Regenerative Medicine Foundation (RMF)** - RMF promotes the field of regenerative medicine through advocacy, education, and research initiatives.

    https://www.regmedfoundation.org/

8. **The Harvard Stem Cell Institute (HSCI)** - HSCI conducts stem cell research and provides resources and training opportunities for scientists in the field.

    https://hsci.harvard.edu/

9. **The New York Stem Cell Foundation (NYSCF)** - NYSCF supports stem cell research through funding, collaboration, and education programs.

    https://nyscf.org/

10. **The Stem Cells Portal** - The Stem Cells Portal provides news, resources, and research updates on stem cell biology and regenerative medicine.

    https://stemcellsportal.com/

Connecting Worlds:

The Evolution of Communication

Communication Technology

Looking ahead to the next century, two groundbreaking technologies stand out: quantum communication and brain-computer interfaces (BCIs).

Quantum communication represents a paradigm shift in how we transmit and secure information. Unlike classical communication systems, which rely on conventional binary encoding and electromagnetic signals, quantum communication harnesses the principles of quantum mechanics to encode information in quantum bits or qubits. These qubits can exist in multiple states simultaneously, enabling secure, high-speed transmission of data over long distances.

One of the most promising applications of quantum communication is quantum key distribution (QKD), a cryptographic protocol that uses quantum principles to establish secure communication channels between parties. QKD leverages the phenomenon of quantum entanglement to generate cryptographic keys that are inherently secure against eavesdropping and interception.

By encoding information in quantum states, QKD ensures that any attempt to intercept or measure the transmitted qubits will disturb their quantum state, alerting the sender and receiver to the presence of an eavesdropper. This level of security is unparalleled in classical encryption methods, making quantum communication a game-changer for industries that require high levels of data privacy and security, such as finance, healthcare, and government.

Furthermore, quantum communication holds the potential to enable quantum teleportation, a phenomenon in which the quantum state of one particle is transferred to another distant particle instantaneously. While the teleportation of macroscopic objects remains the realm of science fiction, quantum teleportation has already been demonstrated at the quantum level, paving the way for future applications in quantum computing, cryptography, and teleportation-based communication networks.

Brain-computer interfaces (BCIs) represent another frontier in communication technology, offering a direct link between the human brain and external devices. BCIs translate neural activity into computer commands or control signals, enabling users to interact with computers, prosthetic devices, and external environments using only their thoughts.

The development of BCIs has been driven by advances in neuroscience, neuroimaging, and machine learning, which have enabled researchers to decode neural signals and extract meaningful information about a user's intentions, movements, and cognitive states. BCIs can be non-invasive, using techniques such as electroencephalography (EEG) to detect electrical signals from the scalp, or invasive, involving the implantation of electrodes directly into the brain tissue.

One of the most promising applications of BCIs is in the field of assistive technology, where they offer new possibilities for individuals with disabilities to regain mobility, communication, and independence. For example, BCIs can enable paralyzed individuals to control robotic limbs, neuroprosthetic devices, or even exoskeletons using their thoughts alone, restoring lost motor function and enhancing quality of life.

BCIs also hold the potential to revolutionize healthcare by enabling new methods of diagnosis, treatment, and rehabilitation for neurological disorders and mental health conditions. Researchers are exploring BCIs for applications such as brain-computer interfaces for stroke rehabilitation, neurofeedback therapy for anxiety and depression, and closed-loop systems for deep brain stimulation in Parkinson's disease and epilepsy.

Furthermore, BCIs have implications for human augmentation and enhancement, blurring the boundaries between humans and machines. As BCIs become more sophisticated and accessible, we may see the emergence of new forms of human-computer interaction, cognitive augmentation, and even telepathic communication, where thoughts and ideas can be transmitted directly between individuals without the need for spoken or written language.

Despite the immense promise of quantum communication and brain-computer interfaces, significant challenges remain on the path to widespread adoption. Technical hurdles such as scalability, reliability, and interoperability must be addressed to realize the full potential of these technologies. Additionally, ethical, legal, and societal considerations surrounding privacy, consent, and access must be carefully navigated to ensure that these technologies are used responsibly and equitably.

Resources

1. **Quantum Information Science and Technology (QIST) at NIST** - NIST conducts research on quantum communication technologies and standards.

    https://www.nist.gov/topics/quantum-information-science

2. **Quantum Communication Hub** - The Quantum Communication Hub is a UK-based initiative focused on advancing research and development in quantum communication technologies.

   https://www.quantumcommshub.net/

3. **Quantum Communication and Cryptography Laboratory at MIT** - MIT's Quantum Communication and Cryptography Laboratory conducts research on quantum communication protocols and applications.

   https://www.rle.mit.edu/qcc/

4. **Center for Quantum Information and Control (CQuIC)** - CQuIC is a research center at the University of New Mexico focused on quantum information science, including quantum communication.

   http://cquic.unm.edu/

5. **Brain-Computer Interface Research at Stanford University** - Stanford's BCI research group conducts research on brain-computer interfaces for applications in healthcare and technology.

   https://bci.stanford.edu/

6. **Brain-Computer Interface Lab at University of Washington** - UW's BCI Lab conducts research on brain-computer interfaces for communication and control applications.

   https://depts.washington.edu/bci/

7. **Center for Neurotechnology (CNT)** - CNT is a research center focused on developing neurotechnologies, including brain-computer interfaces and neural engineering.

   https://www.centerforneurotech.org/

8. **Neurotechnology Center at Columbia University** - Columbia's Neurotechnology Center conducts research on brain-computer interfaces and neural prosthetics.

    https://ntc.columbia.edu/

9. **Brain-Computer Interface Laboratory at University of Essex** - Essex's BCI Lab conducts research on brain-computer interfaces for applications in rehabilitation and assistive technology.

    https://www.essex.ac.uk/bci

10. **Center for Neural Interfaces (CNI) at UC Berkeley** - CNI at UC Berkeley conducts research on neural interfaces, including brain-computer interfaces and neural prosthetics.

    https://neurotech.berkeley.edu/

Data Transmission

As we look towards the future, the evolution of wireless communication systems promises to revolutionize the way we connect, communicate, and interact with the world around us. At the forefront of this evolution is the development of 6G and beyond, ushering in a new era of global connectivity and data transmission.

6G, the sixth generation of wireless communication technology, builds upon the foundation laid by its predecessors, including 5G, 4G, and 3G. While 5G is still in the process of being rolled out globally, researchers and industry leaders are already looking ahead to the next frontier of wireless connectivity. 6G is envisioned to deliver unprecedented speeds, ultra-low latency, and massive device connectivity, unlocking new possibilities for applications such as augmented reality, virtual reality, telemedicine, and autonomous vehicles.

One of the key features of 6G is its ability to operate across a wider range of frequencies, including terahertz (THz) frequencies. Terahertz waves have much shorter wavelengths than microwaves used in previous generations of wireless technology, allowing for higher data transmission rates and faster response times. By harnessing the untapped potential of the terahertz spectrum, 6G promises to deliver multi-gigabit-per-second data rates, enabling real-time communication and immersive experiences with minimal delay.

Moreover, 6G is expected to leverage advanced antenna technologies, such as massive MIMO (Multiple Input Multiple Output) and beamforming, to optimize spectrum efficiency and enhance coverage and capacity. These technologies enable base stations to dynamically adapt their antenna configurations to the surrounding environment, focusing signal energy towards specific users or areas with high demand, while minimizing interference and signal degradation.

Another key aspect of 6G is its support for intelligent networking and self-organizing networks. Machine learning algorithms and artificial intelligence will play a crucial role in optimizing network performance, managing resources, and predicting user behavior. By analyzing vast amounts of data in real-time, 6G networks can dynamically adjust their parameters and configurations to meet the evolving needs of users and applications, ensuring a seamless and reliable connectivity experience.

Furthermore, 6G is expected to enable new forms of communication beyond traditional wireless links, such as wireless sensing and communication through biological tissues. Terahertz waves have the unique ability to penetrate biological tissues without causing harm, making them ideal for applications such as wireless medical diagnostics, neural interfaces, and brain-machine interfaces. By leveraging the biocompatibility of terahertz waves, 6G technology could open up new frontiers in healthcare, enabling non-invasive monitoring and treatment of diseases.

The implications of 6G and beyond for global connectivity and data transmission are profound. By providing ultra-fast, ultra-reliable wireless connectivity, 6G has the potential to bridge the digital divide, connecting underserved communities and remote regions to the global network. This connectivity is not only essential for economic development and social inclusion but also for addressing pressing challenges such as climate change, healthcare access, and education.

Furthermore, 6G technology has the potential to enable new applications and services that were previously unimaginable. From smart cities and connected infrastructure to immersive virtual environments and decentralized networks, the possibilities are endless. 6G could pave the way for a future where everything and everyone is interconnected, ushering in a new era of innovation, collaboration, and human advancement.

However, realizing the full potential of 6G will require concerted efforts from governments, industry stakeholders, and the research community. Investments in research and development, spectrum allocation, and infrastructure deployment will be essential to drive innovation and ensure widespread adoption of 6G technology. Additionally, regulatory frameworks and standards must be established to address issues such as privacy, security, and interoperability, while fostering competition and innovation in the marketplace.

Resources

1. **IEEE Future Networks Initiative** - IEEE's Future Networks Initiative explores emerging technologies, including 6G, and their impact on future wireless communication systems.

    https://futurenetworks.ieee.org/

2. **6G Flagship** - 6G Flagship is a research program in Finland focused on advancing the development of 6G wireless communication technologies.

    https://www.6gflagship.com/

3. **5G and Beyond Wireless Systems (BWS) Lab at NYU** - NYU's BWS Lab conducts research on the evolution of wireless communication systems, including 6G and beyond.

https://wireless.engineering.nyu.edu/research/5g-beyond-wireless-systems-bws-lab/

4. **Wireless Communications and Networking Group (WCNG) at MIT** - MIT's WCNG focuses on research in wireless communication technologies, including 6G and future wireless systems.

   https://wcng.mit.edu/

5. **Centre for Wireless Communications (CWC) at University of Oulu** - CWC at University of Oulu conducts research on wireless communication technologies, including 6G research.

   https://www.oulu.fi/cwc/

6. **Global 6G Alliance (G6GA)** - G6GA is an international alliance focused on driving the development and standardization of 6G technologies.

   https://global6g.org/

7. **5G Innovation Centre (5GIC) at University of Surrey** - 5GIC at University of Surrey conducts research on 5G and beyond wireless communication systems.

   https://www.surrey.ac.uk/5gic

8. **Center for Wireless Communication (CWC) at UC San Diego** - CWC at UC San Diego conducts research on wireless communication technologies, including 6G and beyond.

   https://cwc.ucsd.edu/

9. **Wireless Communications Research Group (WiCR) at Stanford University** - WiCR at Stanford University conducts research on wireless communication technologies, including future wireless systems.

   https://wicr.stanford.edu/

10. **Wireless Communications Lab at Columbia University** - Columbia's Wireless Communications Lab focuses on research in wireless communication technologies, including 6G.

    https://www.ee.columbia.edu/wcl/

Augmented Reality

In the ever-evolving landscape of technology, augmented reality (AR) and virtual reality (VR) stand as transformative innovations with the potential to revolutionize the way we communicate, collaborate, and interact with the world around us. These immersive technologies offer new ways to experience and visualize information, blurring the boundaries between the physical and digital realms. As we look towards the future, the applications of AR and VR in communication and collaboration are poised to reshape industries, education, healthcare, and beyond.

Augmented reality enhances the real-world environment by overlaying digital content, such as images, videos, and 3D models, onto the user's view of the physical world. Unlike virtual reality, which immerses users in entirely virtual environments, AR preserves the user's sense of presence in the real world while augmenting it with digital information. This unique blend of virtual and real-world elements enables a wide range of applications in communication and collaboration.

One of the most promising applications of augmented reality is in remote collaboration and communication. AR technologies enable users to share and interact with digital content in real-time, regardless of their physical location. For example, colleagues working on a project can use AR-enabled smart glasses or mobile devices to view and manipulate 3D models, annotate virtual objects, and communicate with each other as if they were in the same room.

In addition to enhancing remote collaboration, augmented reality has the potential to revolutionize training and education.

By overlaying instructional content onto real-world objects and environments, AR enables hands-on learning experiences that are engaging, interactive, and immersive. For example, medical students can use AR to visualize anatomy in 3D, engineering students can simulate complex machinery, and language learners can practice conversational skills with virtual tutors.

Furthermore, augmented reality has applications in marketing, advertising, and customer engagement. Brands can use AR to create interactive experiences that allow customers to visualize products in their own environments, try on virtual clothing, or preview home furnishings before making a purchase. This immersive shopping experience not only enhances customer engagement but also reduces returns and increases conversion rates.

Virtual reality, on the other hand, immerses users in entirely virtual environments, creating a sense of presence and immersion that is unparalleled by traditional media. VR technologies, such as head-mounted displays and haptic feedback devices, transport users to virtual worlds where they can interact with objects, explore new environments, and engage with other users in real-time.

In the context of communication and collaboration, virtual reality offers new possibilities for remote meetings, training simulations, and virtual events. VR meeting platforms allow users to meet and collaborate in virtual environments, complete with customizable avatars, spatial audio, and interactive whiteboards. These immersive meetings enable more natural communication and collaboration, fostering creativity, innovation, and team cohesion.

Moreover, virtual reality simulations provide a safe and cost-effective way to train for high-risk and complex scenarios, such as medical procedures, emergency response training, and hazardous work environments. By replicating real-world conditions in a virtual environment, VR training simulations enable users to practice skills, make decisions, and learn from mistakes without putting themselves or others at risk.

In the field of healthcare, virtual reality has applications in patient care, medical education, and therapy. VR therapy, for example, has shown promise in treating phobias, post-traumatic stress disorder (PTSD), and chronic pain by immersing patients in virtual environments that expose them to their fears or provide distraction from pain. VR can also be used to simulate medical procedures, allowing students to practice surgery or anatomy dissections in a safe and controlled environment.

Despite their immense potential, augmented reality and virtual reality face challenges in terms of accessibility, usability, and acceptance. High costs, technological limitations, and concerns about privacy and security may hinder widespread adoption of these technologies. Moreover, ethical considerations surrounding the use of AR and VR, such as the potential for addiction, misinformation, and psychological effects, must be carefully addressed to ensure responsible and ethical deployment.

Resources

1. **IEEE Virtual Reality and Augmented Reality Technical Community** - IEEE's Technical Community on Virtual Reality and Augmented Reality provides resources and information on the latest research and developments in VR and AR technologies.

https://tc.computer.org/vr/

2. **XR Association (XRA)** - XRA is an industry consortium focused on promoting the responsible development and adoption of XR technologies, including AR and VR.

   https://www.xrassociation.org/

3. **MIT Media Lab - Responsive Environments Group** - MIT's Responsive Environments Group conducts research on immersive technologies, including AR and VR, for applications in communication, collaboration, and interactive environments.

   https://www.media.mit.edu/groups/responsive-environments/overview/

4. **Stanford Virtual Human Interaction Lab (VHIL)** - Stanford's VHIL conducts research on VR technologies and their applications in communication, collaboration, and social interaction.

   https://vhil.stanford.edu/

5. **Harvard Extended Reality (HXR) Lab** - Harvard's HXR Lab explores the use of extended reality technologies, including AR and VR, in communication, education, and other domains.

   https://hxr.harvard.edu/

6. **Oxford Immersive Technologies and Education Lab (OxITE)** - OxITE at the University of Oxford conducts research on immersive technologies, including AR and VR, for education and communication.

   https://oxite.org/

7. **Association for Computing Machinery (ACM) - SIGGRAPH** - ACM SIGGRAPH focuses on computer graphics and interactive techniques, including VR and AR technologies, for communication and collaboration.

https://www.siggraph.org/

8. **Unity Labs** - Unity Labs, the research division of Unity Technologies, explores the potential of VR and AR technologies for communication, collaboration, and interactive experiences.

   https://unity.com/labs

9. **Immersive Learning Research Network (iLRN)** - iLRN is a global network of researchers and practitioners exploring immersive technologies, including AR and VR, for learning and collaboration.

   https://immersivelrn.org/

10. **Facebook Reality Labs (FRL)** - FRL, formerly Oculus Research, is Facebook's research division focused on advancing VR and AR technologies for communication, social interaction, and collaboration.

    https://www.facebook.com/realitylabs

Eco-Evolution:

Advancing Environmental Innovation

Environmental Challenges

One of the most critical environmental challenges facing humanity today is the rise in atmospheric carbon dioxide levels, primarily driven by the burning of fossil fuels for energy production and transportation. Elevated carbon dioxide levels contribute to global warming, climate change, and ocean acidification, with potentially catastrophic consequences for ecosystems and human societies.

To address this challenge, scientists and engineers are developing carbon capture and storage (CCS) technologies, which aim to capture carbon dioxide emissions from industrial sources such as power plants and factories and store them underground or repurpose them for other uses. CCS technologies encompass a range of approaches, including post-combustion capture, pre-combustion capture, and direct air capture.

Post-combustion capture involves capturing carbon dioxide from the flue gases emitted by power plants and industrial facilities using chemical solvents or adsorbents. Pre-combustion capture involves converting fossil fuels into hydrogen and carbon dioxide before combustion, allowing for the separation and capture of carbon dioxide prior to emission. Direct air capture involves capturing carbon dioxide directly from the atmosphere using chemical processes or sorbent materials.

These carbon capture technologies have the potential to significantly reduce greenhouse gas emissions and mitigate the impacts of climate change. By capturing and sequestering carbon dioxide emissions, CCS technologies can help to stabilize atmospheric carbon dioxide levels and limit global warming, providing a crucial bridge to a low-carbon future while we transition to renewable energy sources.

In addition to carbon capture technologies, sustainable agriculture methods are also playing a vital role in addressing environmental challenges and promoting global food security. Traditional agricultural practices, such as monoculture farming, excessive pesticide use, and soil degradation, have contributed to environmental degradation, loss of biodiversity, and depletion of natural resources.

To address these issues, researchers and farmers are exploring innovative approaches to agriculture that prioritize sustainability, resilience, and biodiversity conservation. These approaches include agroforestry, regenerative agriculture, precision farming, and organic farming, among others.

Agroforestry involves integrating trees and shrubs into agricultural landscapes, providing multiple benefits such as soil conservation, carbon sequestration, and biodiversity enhancement. Regenerative agriculture focuses on restoring and enhancing soil health through practices such as cover cropping, crop rotation, and minimal tillage, which improve soil structure, fertility, and water retention.

Precision farming leverages technology such as GPS, remote sensing, and data analytics to optimize resource use, minimize waste, and increase productivity in agriculture. By precisely monitoring and managing inputs such as water, fertilizers, and pesticides, precision farming can reduce environmental impacts and improve resource efficiency.

Organic farming avoids the use of synthetic pesticides, fertilizers, and genetically modified organisms (GMOs), relying instead on natural methods such as crop rotation, composting, and biological pest control. Organic farming promotes soil health, biodiversity, and ecosystem resilience, while also reducing exposure to harmful chemicals and preserving the integrity of the environment.

These sustainable agriculture methods have the potential to transform the way we produce food, ensuring the long-term viability of agriculture while minimizing its environmental footprint. By promoting biodiversity, soil health, and ecosystem resilience, sustainable agriculture can help to mitigate climate change, protect natural habitats, and ensure food security for future generations.

Resources

1. **Intergovernmental Panel on Climate Change (IPCC)** - The IPCC assesses the scientific basis of climate change, including the impacts of rising carbon dioxide levels.

    https://www.ipcc.ch/

2. **National Aeronautics and Space Administration (NASA)** - Climate Change and Global Warming - NASA provides extensive resources and research findings on climate change, including the role of carbon dioxide emissions.

    https://climate.nasa.gov/

3. **National Oceanic and Atmospheric Administration (NOAA)** - Carbon Tracker - NOAA's Carbon Tracker provides real-time data and research on carbon dioxide concentrations in the atmosphere.

    https://www.esrl.noaa.gov/gmd/ccgg/carbontracker/

4. **Environmental Protection Agency (EPA) - Climate Change** - The EPA offers information on climate change science, impacts, and solutions, including the role of carbon dioxide emissions.

   https://www.epa.gov/climate-change

5. **Union of Concerned Scientists (UCS) - Climate Change** - UCS conducts research and advocacy on climate change, including the impacts of carbon dioxide emissions on the environment and society.

   https://www.ucsusa.org/climate

6. **Carbon Brief** - Carbon Brief provides in-depth analysis and research on the latest developments in climate science, including carbon dioxide emissions and their effects.

   https://www.carbonbrief.org/

7. **World Meteorological Organization (WMO)** - Greenhouse Gas Bulletin - WMO's Greenhouse Gas Bulletin provides annual updates on greenhouse gas concentrations in the atmosphere, including carbon dioxide.

   https://public.wmo.int/en/our-mandate/climate/wmo-greenhouse-gas-bulletin

8. **Global Carbon Project (GCP)** - GCP conducts research on carbon emissions and the global carbon cycle, providing insights into the sources and sinks of carbon dioxide.

   https://www.globalcarbonproject.org/

9. **Center for Climate and Energy Solutions (C2ES)** - C2ES conducts research and provides policy analysis on climate change, including the role of carbon dioxide emissions in driving global warming.

https://www.c2es.org/

10. **The Nature Conservancy - Climate Change** - The Nature Conservancy offers information on climate change impacts and solutions, including strategies for reducing carbon dioxide emissions.

    https://www.nature.org/en-us/what-we-do/our-insights/perspectives/climate-change-perspectives/

The role of biotechnology in environmental conservation and biodiversity preservation

One of the key areas where biotechnology is making a significant impact is in the conservation of endangered species. Many species around the world are facing extinction due to factors such as habitat destruction, poaching, and disease. Biotechnology offers several approaches to address these threats and safeguard the genetic diversity of endangered populations.

One such approach is genetic rescue, which involves using biotechnological tools to increase genetic diversity and reduce the risk of inbreeding in small and isolated populations. Techniques such as artificial insemination, embryo transfer, and genetic rescue through hybridization can help to introduce new genetic variation into endangered populations, enhancing their adaptive potential and resilience to environmental change.

Another promising application of biotechnology in species conservation is genetic engineering, which involves modifying the DNA of organisms to achieve specific conservation goals. For example, scientists are exploring the use of gene editing techniques such as CRISPR-Cas9 to modify the genomes of endangered species to increase their resistance to disease, improve their reproductive success, or enhance their ability to adapt to changing environmental conditions.

Biotechnology also plays a crucial role in monitoring and managing wildlife populations through techniques such as DNA barcoding, which involves using DNA sequences to identify species and track their movements. DNA metabarcoding, a related technique, allows scientists to analyze environmental

DNA (eDNA) samples to detect the presence of species in their habitats, providing valuable data for conservation planning and management.

In addition to species conservation, biotechnology is also being used to restore and rehabilitate degraded ecosystems. One approach is bioremediation, which involves using microorganisms such as bacteria and fungi to degrade pollutants and contaminants in soil, water, and air. Bioremediation can be used to clean up contaminated sites, such as industrial wastelands and oil spills, restoring them to their natural state and promoting ecosystem health.

Another innovative approach to ecosystem restoration is synthetic biology, which involves engineering organisms to perform specific functions or produce valuable compounds. For example, scientists are exploring the use of synthetic biology to engineer plants that can thrive in degraded or contaminated soils, helping to stabilize ecosystems and prevent further erosion and habitat loss.

Biotechnology also offers new opportunities for sustainable agriculture and forestry, which are essential for preserving biodiversity and mitigating climate change. Genetically modified crops, for example, can be engineered to be more resistant to pests and diseases, reducing the need for chemical pesticides and herbicides and minimizing their impact on the environment.

Furthermore, biotechnology can help to improve the productivity and resilience of crops and trees, allowing them to thrive in changing environmental conditions such as drought, heat, and salinity. For example, scientists are developing genetically modified crops that can produce higher yields with less water and fertilizer, helping to conserve resources and reduce the environmental footprint of agriculture.

Resources

1. **Smithsonian Conservation Biology Institute (SCBI)** - SCBI conducts research on wildlife conservation, including the application of biotechnology to protect endangered species.

   https://nationalzoo.si.edu/conservation

2. **IUCN Species Survival Commission (SSC)** - SSC of the International Union for Conservation of Nature (IUCN) focuses on species conservation, including the use of biotechnology for endangered species management.

   https://www.iucn.org/commissions/species-survival-commission

3. **The Wildlife Conservation Society (WCS)** - WCS conducts research and conservation efforts to protect wildlife and wild places, utilizing biotechnology to conserve endangered species.

   https://www.wcs.org/

4. **San Diego Zoo Global - Institute for Conservation Research** - San Diego Zoo Global's Institute for Conservation Research conducts research on conservation genetics and reproductive sciences to aid endangered species recovery.

   https://institute.sandiegozoo.org/

5. **Conservation Genetics Research Group at Cardiff University** - Cardiff University's Conservation Genetics Research Group conducts research on the genetic management of endangered species and the application of biotechnology in conservation.

https://www.cardiff.ac.uk/research/groups-and-centres/conservation-genetics

6. **Centre for Applied Conservation Science at University of Adelaide** - The Centre for Applied Conservation Science at University of Adelaide conducts research on conservation biology, including the use of biotechnology for endangered species conservation.

   https://www.adelaide.edu.au/conservation-science/

7. **Endangered Species Recovery Program at Lincoln Park Zoo** - Lincoln Park Zoo's Endangered Species Recovery Program focuses on conservation breeding and genetic management of endangered species.

   https://www.lpzoo.org/endangered-species-recovery-program

8. **Conservation and Research for Endangered Species (CRES) at San Diego Zoo Global** - CRES conducts research on endangered species conservation, including genetic and reproductive technologies.

   https://cres.sandiegozoo.org/

9. **The Durrell Wildlife Conservation Trust** - The Durrell Wildlife Conservation Trust conducts research and conservation efforts to save species from extinction, utilizing biotechnology when applicable.

   https://www.durrell.org/wildlife/

10. **Conservation Biology Lab at University of Hawai'i at Mānoa** - The Conservation Biology Lab at University of Hawai'i conducts research on wildlife conservation, including genetic approaches to protect endangered species.

    https://www.conservationbiologylab.com/

Geoengineering solutions and their potential impact on climate change mitigation

From solar radiation management to carbon dioxide removal, geoengineering offers a range of potential solutions that could fundamentally alter the trajectory of climate change in the next century.

Solar radiation management (SRM) represents one of the most discussed and debated forms of geoengineering. This approach aims to reduce global temperatures by reflecting a portion of the sun's incoming solar radiation back into space, thereby offsetting the warming effects of greenhouse gases in the atmosphere. One proposed method of SRM involves injecting aerosols, such as sulfur dioxide, into the stratosphere to form a reflective layer that scatters sunlight away from the Earth's surface.

While SRM has the potential to rapidly cool the planet and mitigate the impacts of climate change, it also raises significant ethical, environmental, and geopolitical concerns. Critics argue that SRM could have unintended consequences, such as altering regional weather patterns, disrupting ecosystems, and exacerbating geopolitical tensions over control of the Earth's climate. Furthermore, SRM does not address the root cause of climate change - greenhouse gas emissions - and may merely mask the symptoms without addressing the underlying problem.

Another approach to geoengineering is carbon dioxide removal (CDR), which aims to remove excess carbon dioxide from the atmosphere and store it in long-term reservoirs. CDR techniques include afforestation and reforestation, which involve planting trees and restoring forests to absorb carbon dioxide through photosynthesis, as well as direct air capture (DAC), which involves capturing carbon dioxide directly from the atmosphere using chemical processes or sorbent materials.

While CDR has the potential to reverse the accumulation of carbon dioxide in the atmosphere and mitigate the long-term impacts of climate change, it also presents challenges such as land use competition, energy requirements, and cost-effectiveness. Afforestation and reforestation projects require vast areas of land and may compete with food production and biodiversity conservation. DAC technologies are energy-intensive and may require significant investment in infrastructure and research and development to achieve scalability and cost-effectiveness.

In addition to SRM and CDR, other geoengineering approaches are being explored, such as ocean fertilization, cloud seeding, and enhanced weathering. Ocean fertilization involves adding nutrients to the ocean to stimulate phytoplankton growth, which can absorb carbon dioxide from the atmosphere through photosynthesis. Cloud seeding involves seeding clouds with particles to enhance cloud formation and increase their reflectivity, while enhanced weathering involves accelerating the natural weathering process of rocks to absorb carbon dioxide from the atmosphere.

While these geoengineering solutions hold promise for mitigating climate change, they also raise ethical, environmental, and regulatory concerns. Ocean fertilization could disrupt marine ecosystems and lead to unintended consequences such as harmful algal blooms or ocean acidification. Cloud seeding and enhanced weathering could have unpredictable impacts on regional climate patterns and precipitation, with potentially far-reaching consequences for agriculture, water resources, and ecosystems.

Furthermore, the deployment of geoengineering technologies could exacerbate existing inequalities and geopolitical tensions, as different countries may have divergent interests and priorities regarding their use. International cooperation and governance frameworks will be essential to ensure that geoengineering is deployed responsibly, transparently, and equitably, with due consideration given to the potential risks, benefits, and unintended consequences.

Resources

1. **Harvard Solar Geoengineering Research Program (SGRP)** - SGRP conducts research on solar radiation management and other geoengineering approaches to mitigate climate change.

    https://geoengineering.environment.harvard.edu/

2. **Oxford Geoengineering Programme** - The Oxford Geoengineering Programme conducts research on the science, ethics, and governance of geoengineering solutions for climate change.

    https://www.geoengineering.ox.ac.uk/

3. **Carnegie Climate Geoengineering Governance Initiative (C2G2)** - C2G2 focuses on governance and public engagement related to geoengineering research and deployment.

    https://www.c2g2.net/

4. **Geoengineering Our Climate (GEOCLIM)** - GEOCLIM is a European research project exploring the potential risks and benefits of geoengineering approaches to climate change mitigation.

    https://www.geoclim.eu/

5. **National Center for Atmospheric Research (NCAR)** - Geoengineering Research - NCAR conducts research on geoengineering approaches, including modeling studies and climate simulations.

    https://ncar.ucar.edu/research/theme/geoengineering

6. **Centre for International Governance Innovation (CIGI)** - Geoengineering Governance Initiative - CIGI's Geoengineering Governance Initiative focuses on the governance challenges associated with geoengineering research and deployment.

    https://www.cigionline.org/project/geoengineering-governance

7. **Geoengineering Model Intercomparison Project (GeoMIP)** - GeoMIP coordinates climate model simulations to assess the potential impacts of geoengineering interventions on the climate system.

    https://www.geosociety.org/GSA/GeoMIP/

8. **Geoengineering Watch** - Geoengineering Watch provides news, analysis, and advocacy on geoengineering research, governance, and potential impacts.

    https://www.geoengineeringwatch.org/

9. **Royal Society** - Solar Radiation Management Governance Initiative (SRMGI) - SRMGI focuses on governance and public engagement related to solar radiation management research and potential deployment.

    https://www.srmgi.org/

10. **Geoengineering Policy Observatory** - The Geoengineering Policy Observatory provides resources

and analysis on the policy, governance, and ethical dimensions of geoengineering solutions.

https://geoengineeringpolicy.org/

Frontiers Beyond:

Charting the Course of Space Exploration and Colonization

Space Exploration Technology

One of the most significant breakthroughs in space exploration technology is the development of reusable rockets. Traditionally, rockets have been single-use vehicles, with each launch requiring the construction of a new rocket from scratch. This approach is not only costly but also wasteful, resulting in the loss of valuable resources and limiting the frequency and accessibility of space missions.

Reusable rockets, on the other hand, are designed to be launched, landed, and refurbished for multiple missions, significantly reducing the cost and complexity of space travel. Companies such as SpaceX, Blue Origin, and Rocket Lab have pioneered the development of reusable rocket technology, successfully demonstrating the feasibility of landing and reusing rocket stages to lower the cost of access to space.

The advent of reusable rockets has opened up new possibilities for space exploration, enabling more frequent and affordable missions to Earth orbit, the Moon, Mars, and beyond. With reusable rockets, space agencies and private companies can conduct scientific research, deploy satellites, and build infrastructure in space more efficiently, paving the way for the exploration and colonization of other worlds.

Another exciting frontier in space exploration technology is asteroid mining, which involves extracting valuable resources such as water, metals, and rare minerals from asteroids and other celestial bodies. Asteroids are rich in resources that are scarce or expensive to obtain on Earth, including platinum, gold, and water ice, which can be used to support future space missions and sustain human settlements in space.

Asteroid mining has the potential to revolutionize the space industry by providing a sustainable source of raw materials for spacecraft construction, fuel production, and life support systems. Water ice, in particular, is highly valuable as it can be converted into hydrogen and oxygen for rocket propulsion and life support, enabling long-duration missions and refueling operations in space.

Furthermore, asteroid mining could lead to the development of new technologies and industries, creating jobs and economic opportunities both in space and on Earth. Companies such as Planetary Resources and Deep Space Industries are actively exploring the feasibility of asteroid mining and developing the necessary technologies to extract and process resources in space.

In addition to reusable rockets and asteroid mining, other advancements in space exploration technology are also shaping the future of space exploration. For example, advancements in propulsion systems, such as ion engines and nuclear propulsion, could enable faster and more efficient travel to distant destinations within our solar system and beyond.

Moreover, advances in robotics, artificial intelligence, and autonomous systems are revolutionizing the way we explore and study space. Robotic spacecraft such as NASA's Perseverance rover and the European Space Agency's Rosetta mission have demonstrated the capabilities of autonomous navigation, sample collection, and scientific analysis, paving the way for future robotic missions to explore the far reaches of the solar system.

Furthermore, the emergence of commercial spaceflight companies such as SpaceX, Blue Origin, and Virgin Galactic is democratizing access to space and fostering innovation in the space industry. These companies are developing new technologies, business models, and markets for space tourism, satellite deployment, and beyond-Earth activities, opening up new opportunities for collaboration and exploration.

Resources

1. **SpaceX** - SpaceX is a leading aerospace manufacturer and space transportation company that has pioneered the development of reusable rocket technology.

   https://www.spacex.com/

2. **Blue Origin** - Blue Origin, founded by Amazon's Jeff Bezos, is another prominent company working on reusable rocket technology and space exploration.

   https://www.blueorigin.com/

3. **NASA - Space Technology Mission Directorate (STMD)** - NASA's STMD invests in innovative technologies, including reusable rocket systems, to enable future space exploration missions.

   https://www.nasa.gov/directorates/spacetech/home

4. **European Space Agency (ESA) - Future Launchers Preparatory Programme (FLPP)** - ESA's FLPP focuses on developing next-generation launch systems, including reusable rocket technologies.

   https://www.esa.int/ESA_Multimedia/Images/2016/05/Future_Launchers_Preparatory_Programme

5. **Rocket Lab** - Rocket Lab is a private aerospace company that specializes in small satellite launch services and is also exploring reusable rocket technology.

   https://www.rocketlabusa.com/

6. **United Launch Alliance (ULA)** - ULA is a joint venture between Boeing and Lockheed Martin that provides launch services using expendable rockets, but they are also exploring reusable rocket concepts.

    https://www.ulalaunch.com/

7. **Virgin Galactic -** Virgin Galactic, founded by Richard Branson, is known for its suborbital space tourism flights but is also developing reusable rocket technology for orbital launches.

    https://www.virgingalactic.com/

8. **Relativity Space** - Relativity Space is a startup that aims to revolutionize rocket manufacturing and launch with 3D printing technology and reusable rockets.

    https://www.relativityspace.com/

9. **Arianespace** - Arianespace is a European launch service provider that primarily uses expendable rockets but is also exploring options for reusable launch vehicles.

    https://www.arianespace.com/

10. **Jeff Bezos' 2019 Washington Post interview about Blue Origin's vision for the future** - This interview provides insights into Blue Origin's approach to space exploration technology, including reusable rockets.

    https://www.washingtonpost.com/technology/2019/04/05/jeff-bezos-unveils-blue-origins-vision-space-exploration/

Human Colonization of Mars

Mars, often referred to as the "red planet," has long been a focal point of human exploration and curiosity. With its relatively Earth-like climate, abundant resources, and proximity to Earth, Mars has emerged as one of the most promising candidates for human colonization within our solar system. The idea of sending humans to Mars has been discussed and debated for decades, with numerous space agencies and private companies outlining ambitious plans for crewed missions to the red planet.

One of the key challenges of human colonization of Mars is the vast distance between Earth and Mars, which can take anywhere from six to nine months to travel one way using current propulsion technologies. This long-duration space travel presents numerous logistical, physiological, and psychological challenges for astronauts, including radiation exposure, muscle atrophy, bone density loss, and psychological stress. To address these challenges, researchers and engineers are developing advanced spacecraft, habitats, and life support systems capable of sustaining human life during the long journey to Mars and beyond.

Once humans arrive on Mars, they will face a host of challenges associated with living and working in a harsh and inhospitable environment. Mars has a thin atmosphere composed primarily of carbon dioxide, with surface temperatures averaging around minus 80 degrees Fahrenheit (-62 degrees Celsius) and frequent dust storms that can obscure visibility and damage equipment. Furthermore, Mars lacks liquid water on its surface, making it necessary to find alternative sources of water for drinking, irrigation, and industrial purposes.

Despite these challenges, Mars offers several advantages for human colonization, including abundant resources such as water ice, carbon dioxide, and minerals that could be extracted and utilized to support human settlements. Water ice, in particular, is highly valuable as it can be melted and purified to provide drinking water, support agriculture, and generate oxygen and hydrogen for life support and rocket fuel production. In addition, the Martian soil contains nutrients that could be used to grow crops and support local ecosystems in greenhouses or controlled environments.

Furthermore, Mars has a day-night cycle and a tilt similar to Earth's, which could facilitate the development of renewable energy sources such as solar power and wind power to meet the energy needs of human settlements. Solar panels could be deployed on the surface of Mars to capture sunlight and generate electricity, while wind turbines could harness the Martian atmosphere to generate power.

In addition to Mars, other celestial bodies within our solar system, such as the Moon, asteroids, and moons of gas giants like Jupiter and Saturn, also hold potential for human colonization. The Moon, in particular, has attracted attention as a possible stepping stone for future human missions to Mars and beyond. With its proximity to Earth and abundant resources such as water ice at the lunar poles, the Moon could serve as a testbed for developing and testing technologies and techniques for long-duration space missions and human settlement.

Furthermore, the Moon's low gravity and lack of atmosphere make it an ideal location for launching spacecraft and exploring the outer solar system and beyond. By establishing a permanent presence on the Moon, humans could develop infrastructure and capabilities for deep-space exploration and colonization, paving the way for future missions to Mars, asteroids, and other destinations in the solar system and beyond.

Resources

1. **NASA - Mars Exploration Program** - NASA's Mars Exploration Program provides comprehensive information on past, present, and future missions to Mars, including plans for human exploration and colonization.

    https://mars.nasa.gov/

2. **SpaceX - Mars** - SpaceX, founded by Elon Musk, has outlined ambitious plans for human colonization of Mars, including the development of the Starship spacecraft.

    https://www.spacex.com/vehicles/starship/

3. **The Mars Society** - The Mars Society is a nonprofit organization dedicated to promoting human exploration and colonization of Mars through research, advocacy, and public outreach.

    https://www.marssociety.org/

4. **European Space Agency (ESA) - ExoMars** - ESA's ExoMars program includes missions to study the Martian environment and potential resources, laying the groundwork for future human exploration.

    https://www.esa.int/Science_Exploration/Human_and_Robotic_Exploration/Exploration/ExoMars

5. **Mars One** - Mars One was a nonprofit organization that aimed to establish a permanent human settlement on Mars, although the project has faced challenges and setbacks.

    https://www.mars-one.com/

6. **National Geographic - Mars Exploration** - National Geographic provides articles, videos, and documentaries on Mars exploration and the potential for human colonization.

   https://www.nationalgeographic.com/magazine/2013/05/mars-exploration/

7. **The Planetary Society - Humans on Mars** - The Planetary Society advocates for human exploration of Mars and provides resources on the scientific, technical, and ethical aspects of colonization.

   https://www.planetary.org/planetary-defense/humans-on-mars

8. **Space.com - Mars** - Space.com offers news, articles, and features on Mars exploration, including plans for human missions and colonization.

   https://www.space.com/topics/mars

9. **NASA Astrobiology Institute - Mars** - NASA's Astrobiology Institute conducts research on the potential for life on Mars and the implications for human colonization.

   https://astrobiology.nasa.gov/research/astrobiology-at-nasa/mars/

10. **University of Arizona - Lunar and Planetary Laboratory (LPL)** - LPL conducts research on planetary science, including Mars exploration and the potential for human habitation.

    https://www.lpl.arizona.edu/

Space Tourism

Space tourism, the practice of traveling to space for recreational, leisure, or adventure purposes, traces its origins back to the pioneering days of human spaceflight. The first space tourists were the astronauts of the Apollo program, who ventured to the Moon in the 1960s and 1970s as part of NASA's efforts to explore and conquer the final frontier. However, it was not until the late 20th and early 21st centuries that space tourism began to take shape as a commercial industry.

In 2001, the world witnessed the historic launch of Dennis Tito, an American businessman and former NASA engineer, as the first space tourist to visit the International Space Station (ISS) aboard a Russian Soyuz spacecraft. Tito's mission marked the beginning of a new era in space exploration, opening the door for private individuals to travel to space and experience the awe-inspiring beauty and majesty of the cosmos firsthand.

Since then, several companies have emerged as leaders in the burgeoning space tourism industry, offering a range of experiences and services for aspiring space travelers. Companies such as SpaceX, Blue Origin, Virgin Galactic, and Space Adventures are developing spacecraft, launch vehicles, and space habitats designed to transport tourists to destinations such as Earth orbit, the Moon, and even beyond.

One of the most promising developments in space tourism is the advent of suborbital spaceflight, which offers brief but exhilarating journeys to the edge of space and back. Companies like Virgin Galactic and Blue Origin are developing spacecraft that can carry passengers on suborbital flights, allowing them to experience weightlessness and see the curvature of the Earth against the backdrop of the cosmos.

Suborbital space tourism holds the potential to revolutionize the way we think about space travel, offering a more accessible, affordable, and immersive experience for people from all walks of life. Unlike traditional space missions, which require months of training and preparation, suborbital spaceflights can be completed in a matter of days or even hours, making them accessible to a wider range of participants, including tourists, scientists, educators, and artists.

Furthermore, suborbital space tourism could pave the way for future missions to more distant destinations, such as the Moon, Mars, and beyond. By demonstrating the feasibility and safety of commercial space travel, suborbital spaceflights could stimulate investment, innovation, and collaboration in the space industry, accelerating progress towards the next frontier of human exploration and colonization.

In addition to suborbital space tourism, companies like SpaceX and Space Adventures are developing missions that will take tourists on orbital flights around the Earth and beyond. These missions offer longer-duration experiences that allow passengers to live and work in space, conduct scientific research, and participate in extravehicular activities (EVAs) such as spacewalks.

Orbital space tourism represents the pinnacle of human spaceflight, offering an unparalleled opportunity to experience the wonders of space travel and the beauty of our planet from a perspective that few have ever witnessed. While orbital spaceflights require more extensive training and preparation than suborbital flights, they offer a truly transformative experience that can inspire and empower individuals to see themselves as citizens of the cosmos.

In addition to its recreational and inspirational value, space tourism also holds the potential to drive economic growth, stimulate technological innovation, and expand our understanding of the universe. The space tourism industry has already created thousands of jobs and generated billions of dollars in revenue, supporting a wide range of businesses and industries, including aerospace, hospitality, entertainment, and tourism.

Furthermore, space tourism has the potential to catalyze breakthroughs in space exploration technology, infrastructure, and capabilities, which can benefit not only tourists but also astronauts, scientists, engineers, and researchers. By investing in space tourism, companies and governments can accelerate progress towards ambitious goals such as lunar exploration, Mars colonization, and the search for extraterrestrial life.

Resources

1. **Virgin Galactic** - Virgin Galactic is a prominent company in the space tourism industry, offering suborbital spaceflight experiences for private individuals.

    https://www.virgingalactic.com/

2. **Blue Origin** - Blue Origin, founded by Jeff Bezos, is another key player in the space tourism sector, with plans to offer suborbital flights aboard its New Shepard spacecraft.

    https://www.blueorigin.com/

3. **Space Adventures** - Space Adventures is a space tourism company that has facilitated orbital

spaceflights for private individuals to visit the International Space Station (ISS) aboard Russian Soyuz spacecraft.

https://www.spaceadventures.com/

4. **NASA** - Commercial Crew Program - NASA's Commercial Crew Program works with private companies to develop spacecraft capable of carrying astronauts to and from the ISS, with potential applications for space tourism.

https://www.nasa.gov/exploration/commercial/crew/index.html

5. **Axiom Space** - Axiom Space is a company working to develop commercial space stations and space tourism opportunities, including missions to the ISS.

https://www.axiomspace.com/

6. **SpaceX - Inspiration4** - SpaceX's Inspiration4 mission, the world's first all-civilian spaceflight, marked a significant milestone in the advancement of space tourism.

https://www.inspiration4.com/

7. SpaceX - Starship - SpaceX's Starship spacecraft is designed for a variety of missions, including space tourism, with plans for lunar and Mars exploration as well.

https://www.spacex.com/vehicles/starship/

8. **Space Tourism Society** - The Space Tourism Society is a nonprofit organization dedicated to promoting space tourism and facilitating collaboration within the industry.

https://spacetourismsociety.org/

9. **NASA - Tourism on the ISS** - NASA provides information on the potential for space tourism to the ISS, including recent developments and opportunities.

   https://www.nasa.gov/mission_pages/station/tourism/

10. **Astro Tourism - International Dark-Sky Association** - While not focused solely on space tourism, the International Dark-Sky Association promotes astronomy-based tourism experiences, including stargazing and space observation.

    https://www.darksky.org/

Conclusion

From renewable energy and transportation to healthcare, communication, and space exploration, the inventions featured in this book represent the collective ingenuity, creativity, and determination of humanity to overcome challenges, seize opportunities, and push the boundaries of what is possible. Each invention offers a glimpse into a future where innovation drives progress, where technology serves humanity, and where the impossible becomes possible.

As we reflect on the implications of these inventions, it is essential to recognize the profound impact they may have on our lives, our societies, and our planet. The transition to renewable energy sources promises to mitigate the effects of climate change and ensure a sustainable future for generations to come. Breakthroughs in healthcare offer the hope of curing diseases, prolonging life, and enhancing human well-being. Advancements in communication technology connect us in new and unprecedented ways, fostering collaboration, understanding, and empathy across borders and cultures. Moreover, the exploration of space holds the promise of unlocking the mysteries of the cosmos and expanding the horizons of human knowledge and understanding. As we venture beyond Earth to explore distant planets, asteroids, and galaxies, we gain a deeper appreciation for the interconnectedness of all life and the fragility of our planet.

As we look ahead to the next 100 years, it is clear that the future is ours to shape and define. The inventions and technologies featured in this book offer a roadmap for progress, a blueprint for innovation, and a call to action for all those who dare to dream, dare to innovate, and dare to imagine a world where anything is possible.

www.ingramcontent.com/pod-product-compliance
Lightning Source LLC
Chambersburg PA
CBHW071057240526
45471CB00016B/1980